ANIMALS OF THE SERENGETI
Wildlife of East Africa Encyclopedias for Children

SPEEDY PUBLISHING

Speedy Publishing LLC
40 E. Main St. #1156
Newark, DE 19711
www.speedypublishing.com

Copyright 2015

All Rights reserved. No part of this book may be reproduced or used in any way or form or by any means whether electronic or mechanical, this means that you cannot record or photocopy any material ideas or tips that are provided in this book

The Serengeti is a region of savannah in East Africa.

Elephants are the largest land-living mammal in the world.

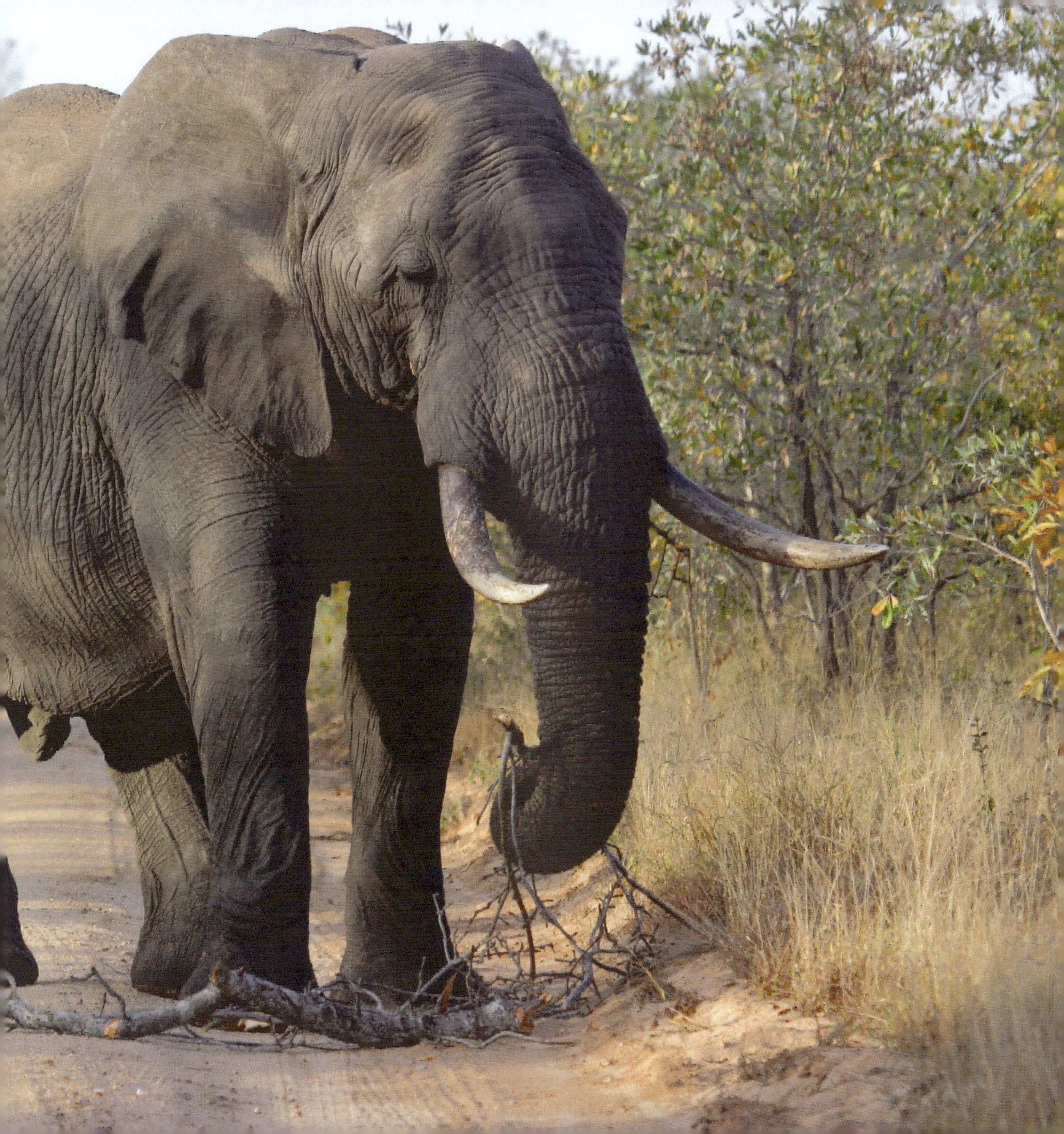

Elephants are herbivores meaning they only eat plants and vegetables. Elephants travel in a herd lead by a female.

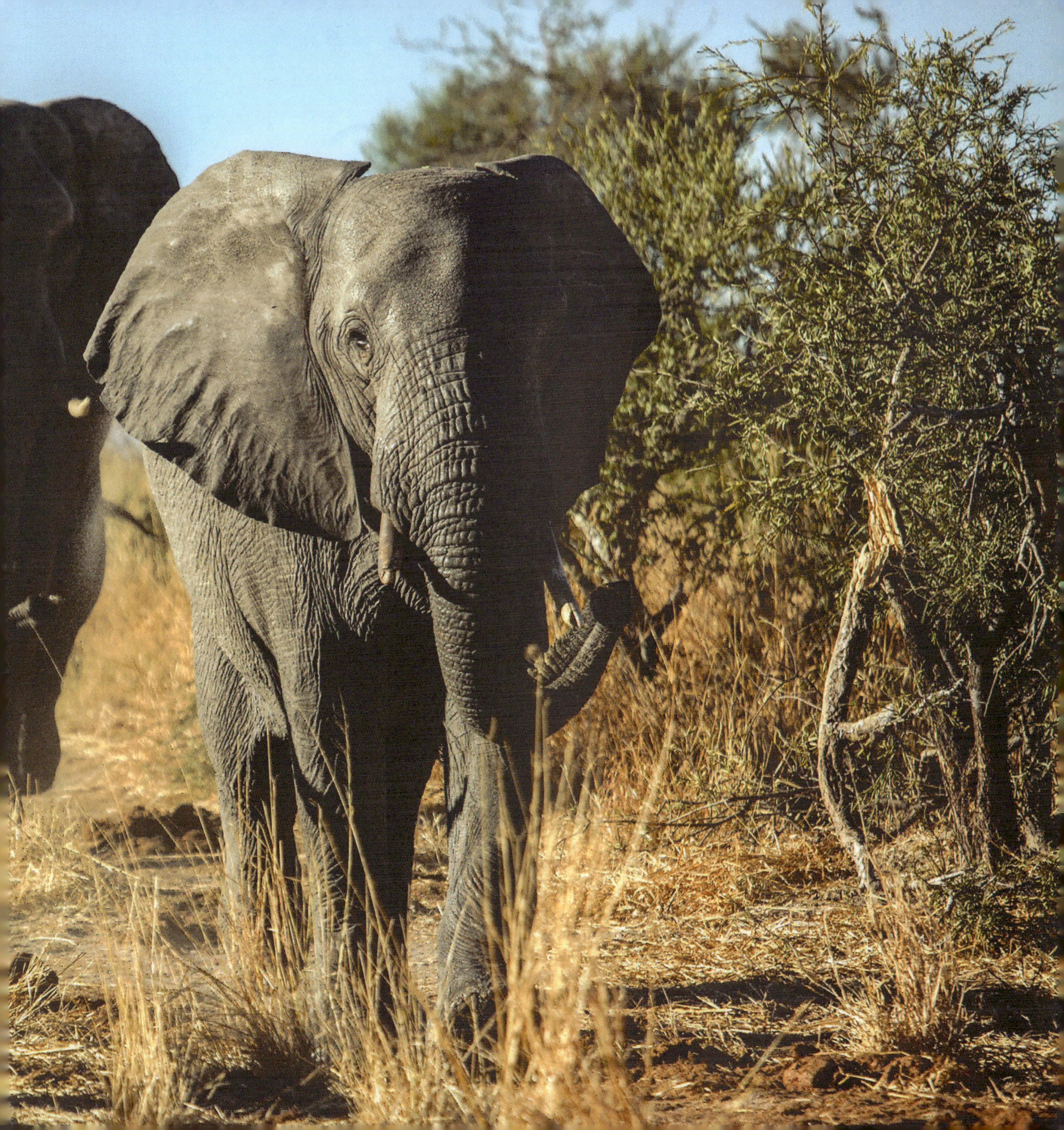

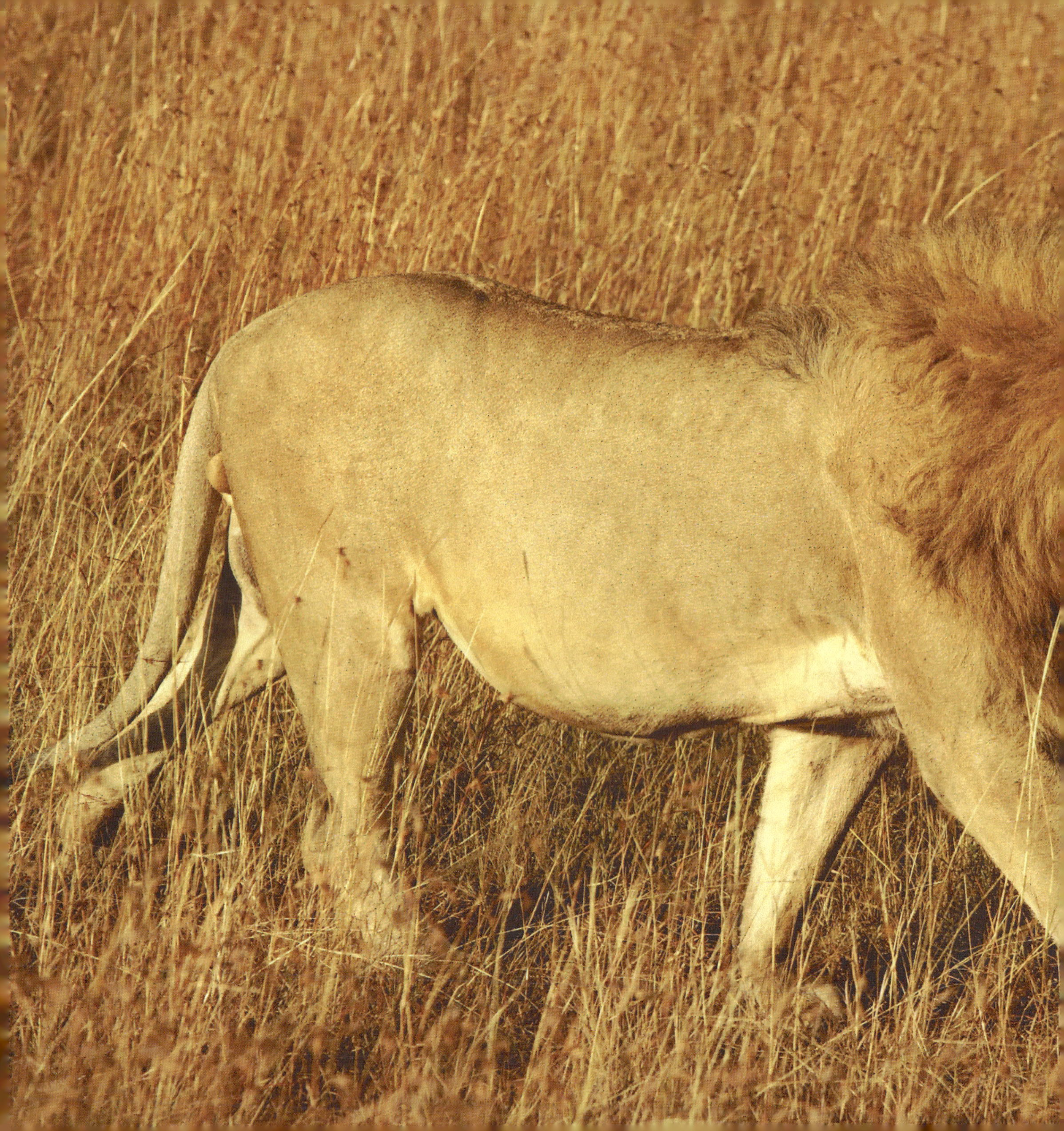

Lions are the second largest big cat species in the world.

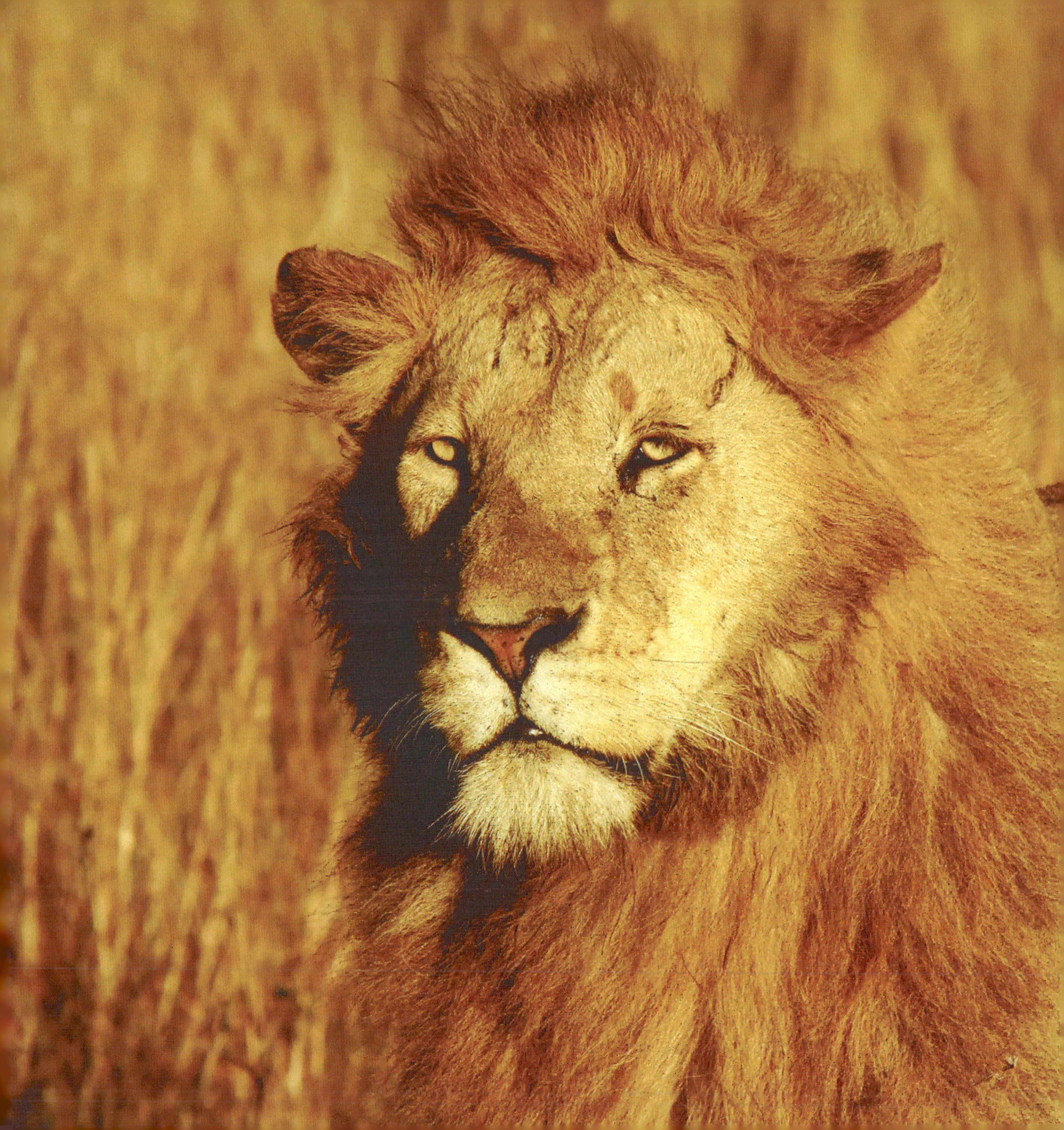

Lion's are known as king of the jungle, but they really don't live in jungles.

Hippopotamuses are a semi-aquatic animal meaning they spend a lot of the time in the water.

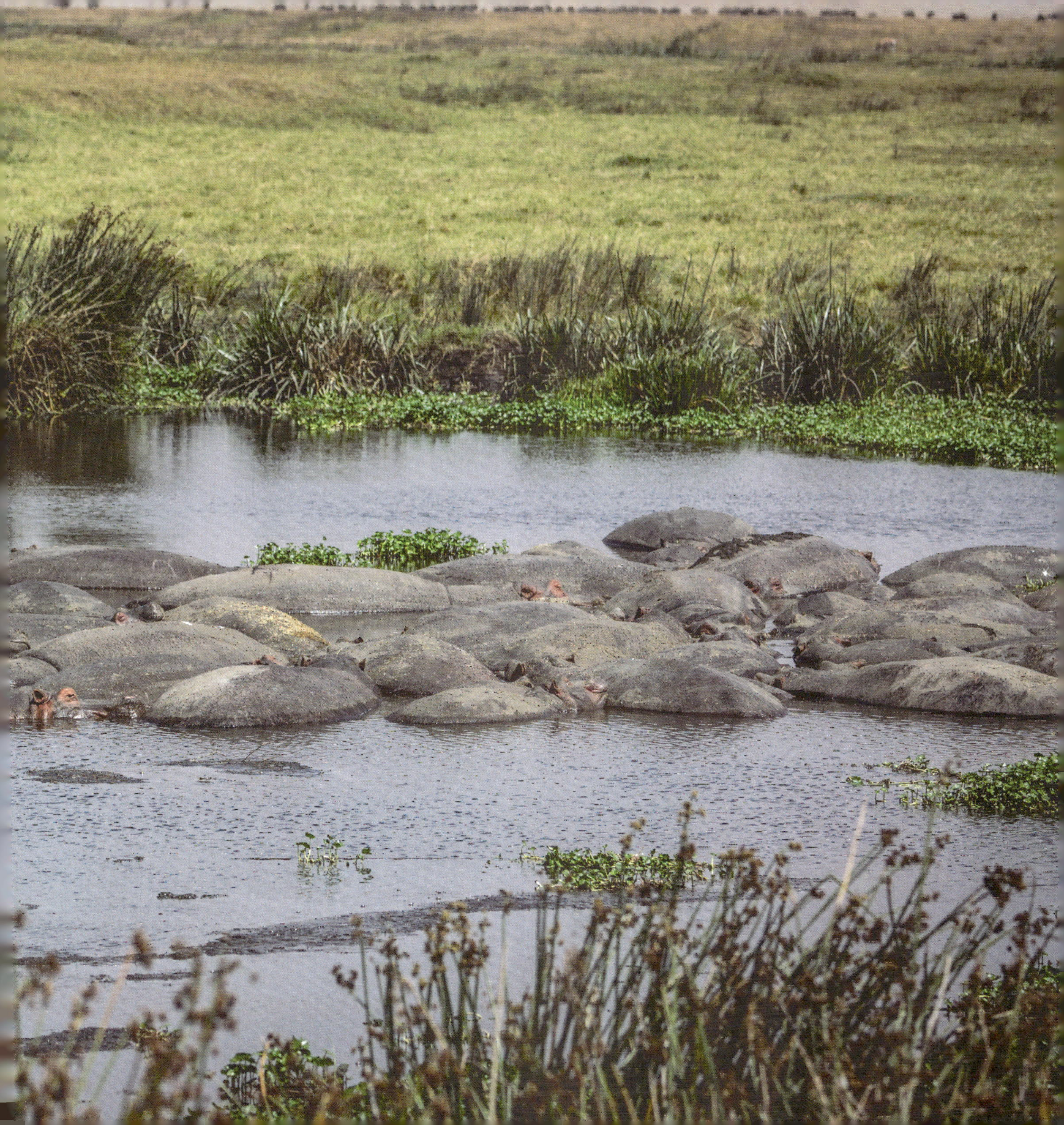

They are regarded as one of the most dangerous animals in Africa.

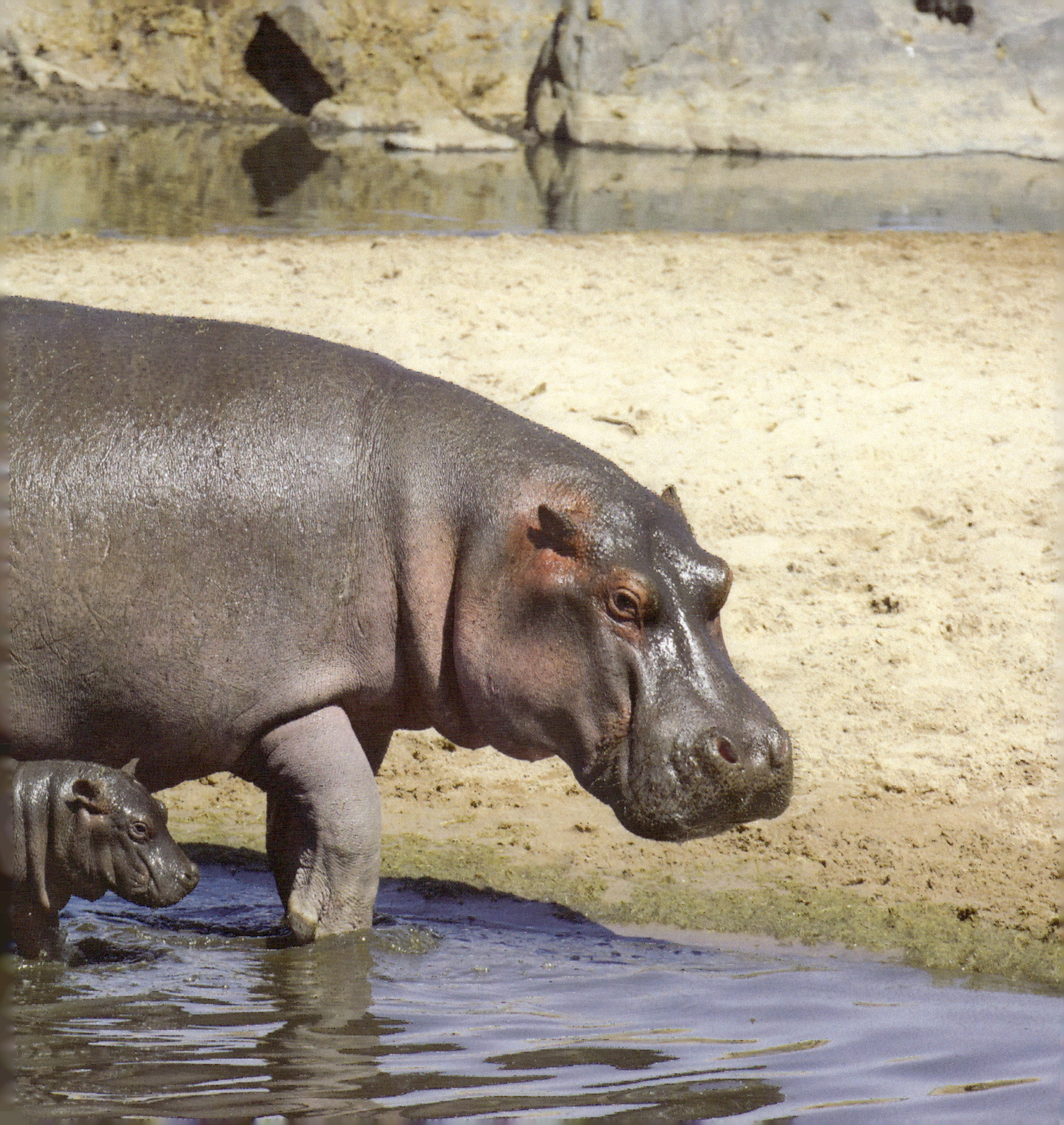

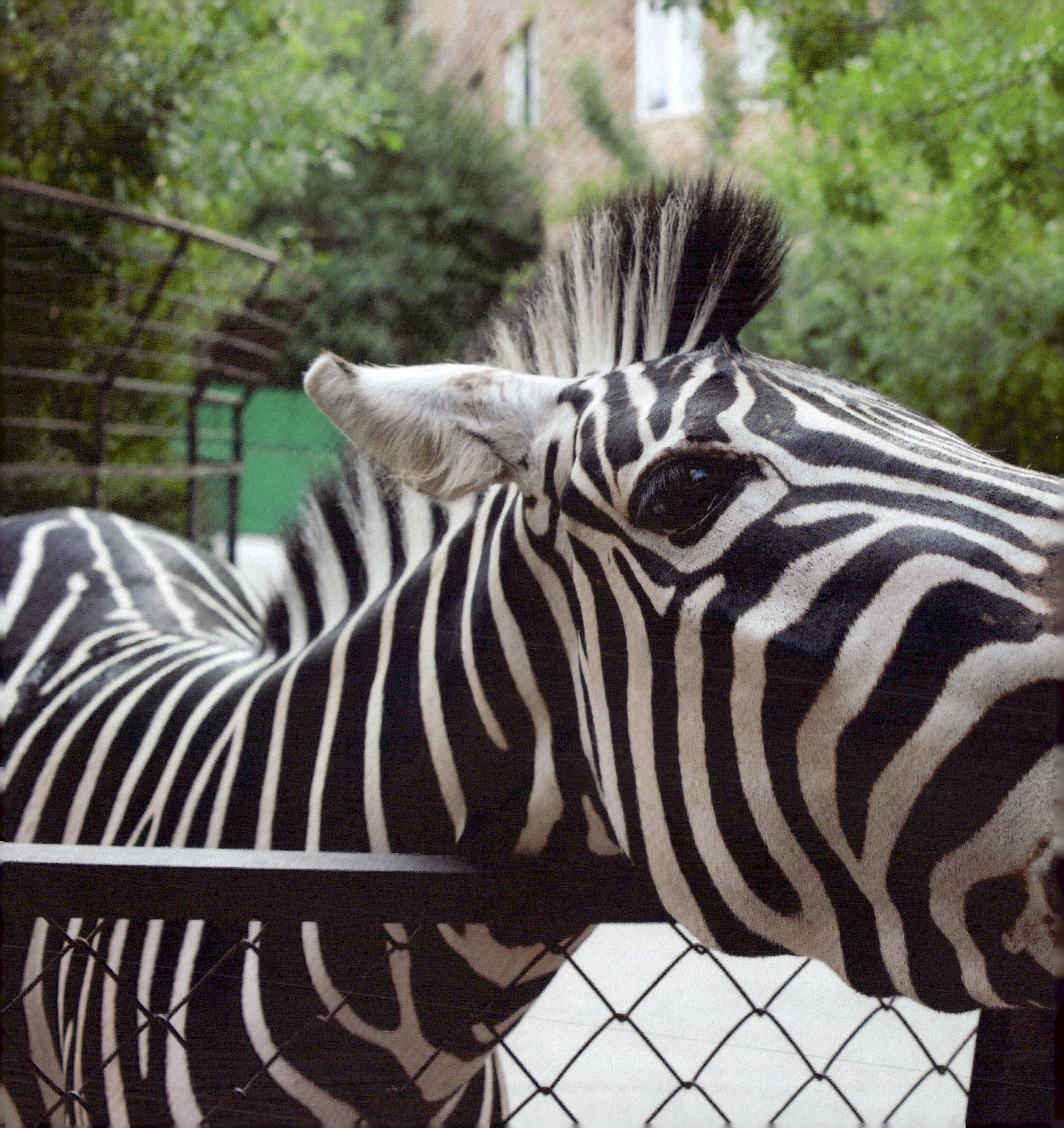

Zebras are members of the horse family. Every zebra has a unique pattern of black and white stripes.

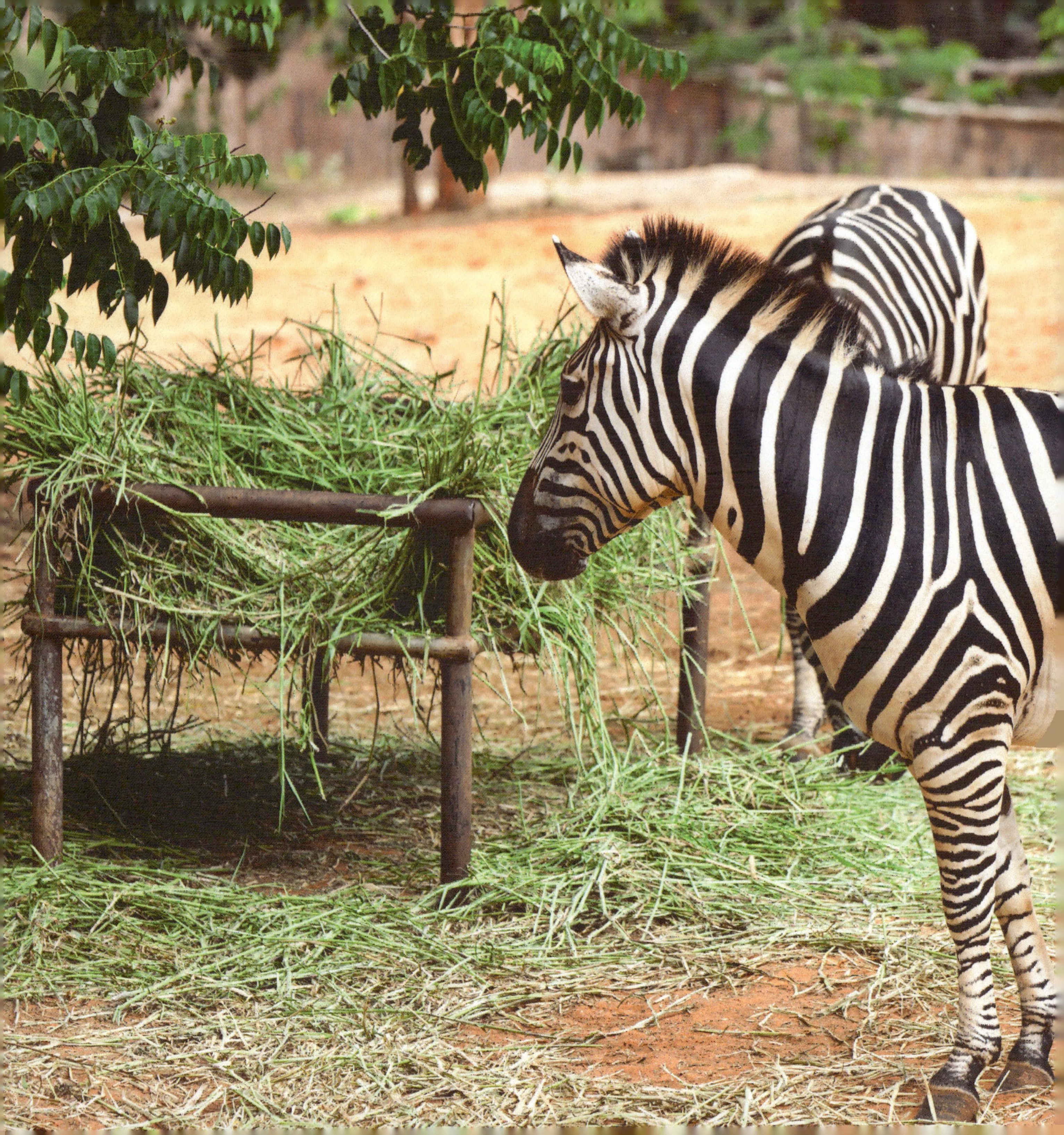

When zebras are grouped together, their stripes make it hard for a lion or leopard to pick out one zebra to chase.

Giraffes are the tallest land animals. A giraffe can live 25 years in the wild and 30 in captivity.

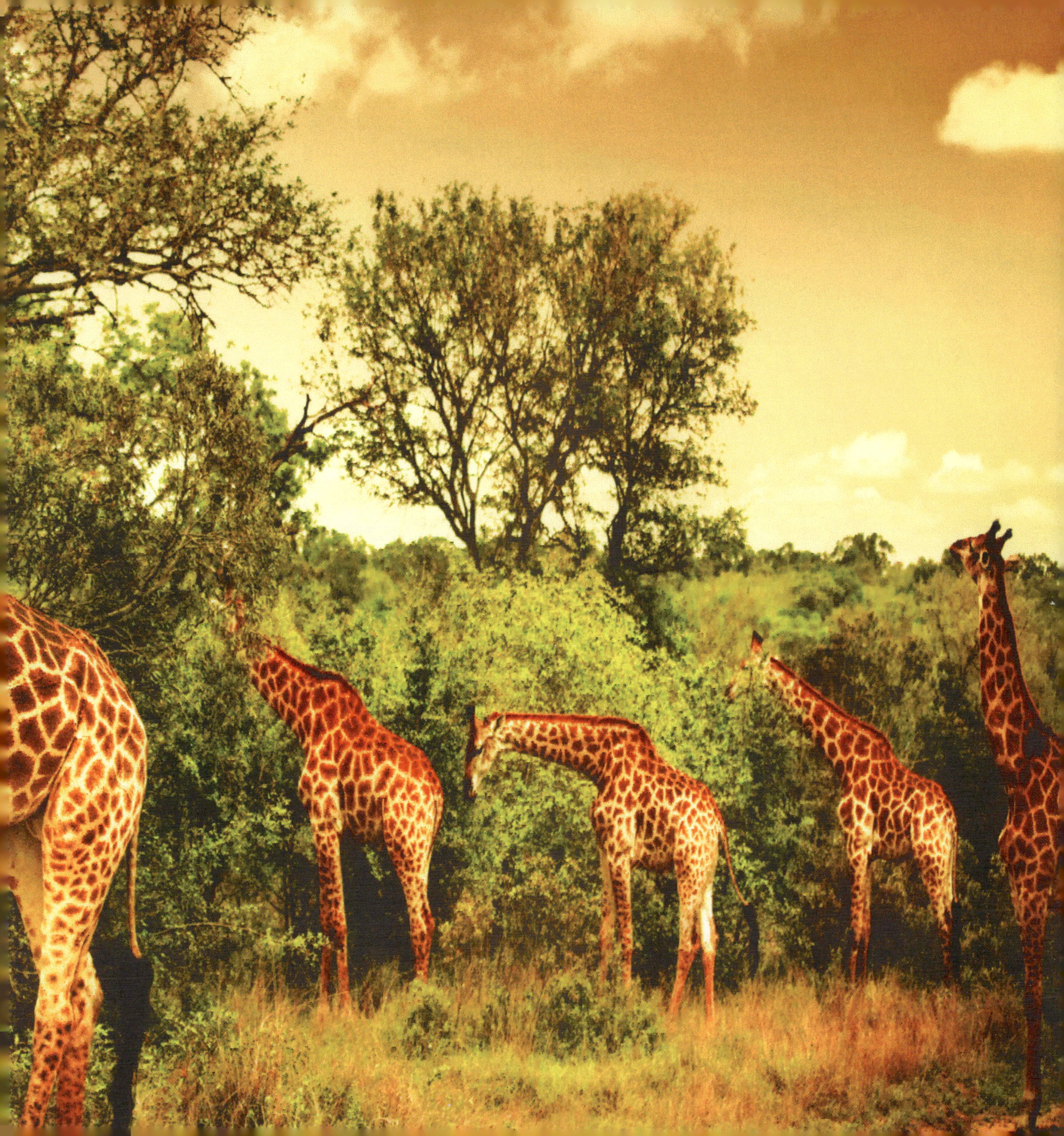

Giraffes eat leaves and branches mostly from acacia, mimosa and wild apricot trees.

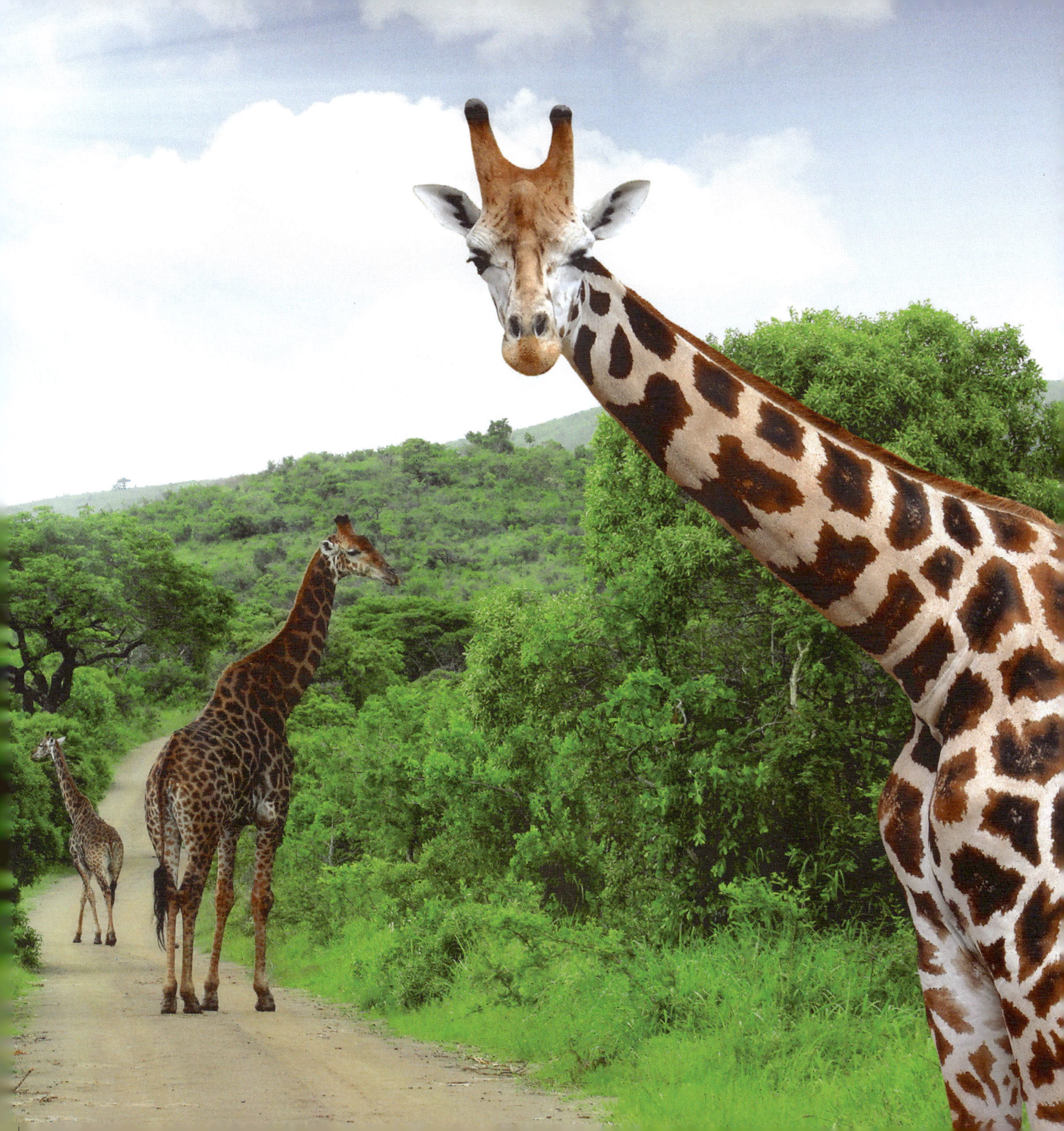

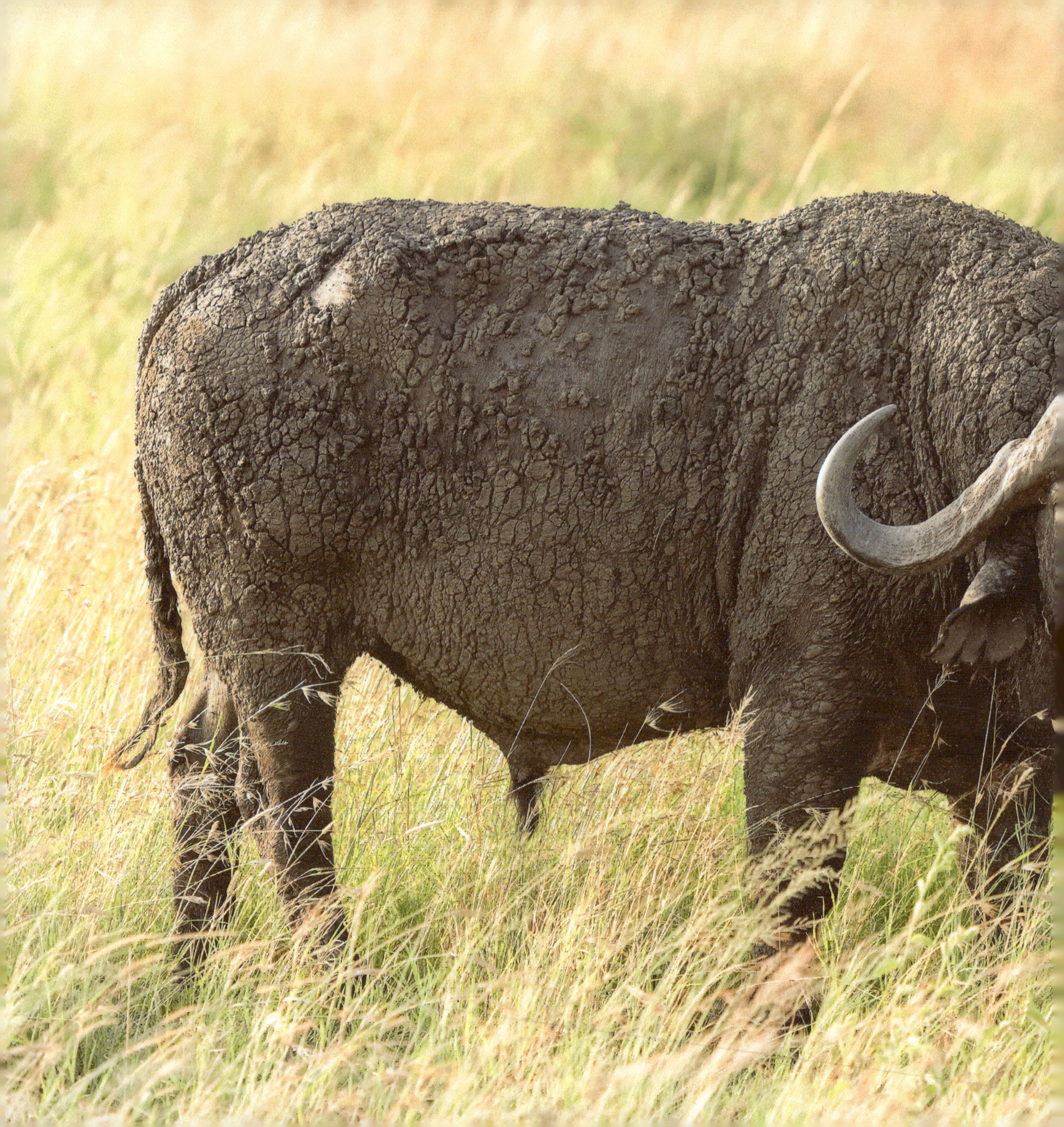

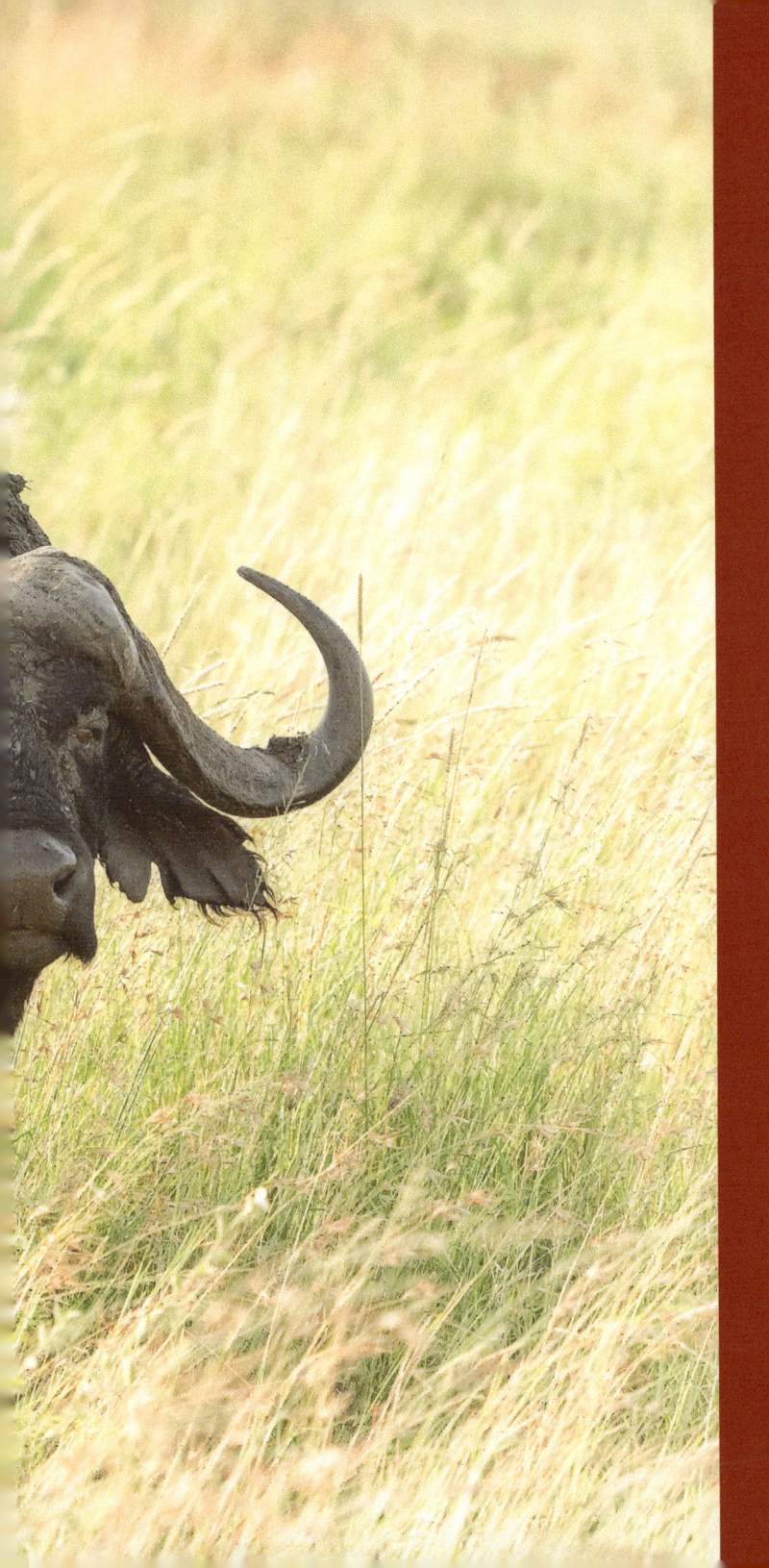

African buffalo is a large animal that can reach 6.8 to 11 feet in length and weigh between 660 and 1900 pounds.

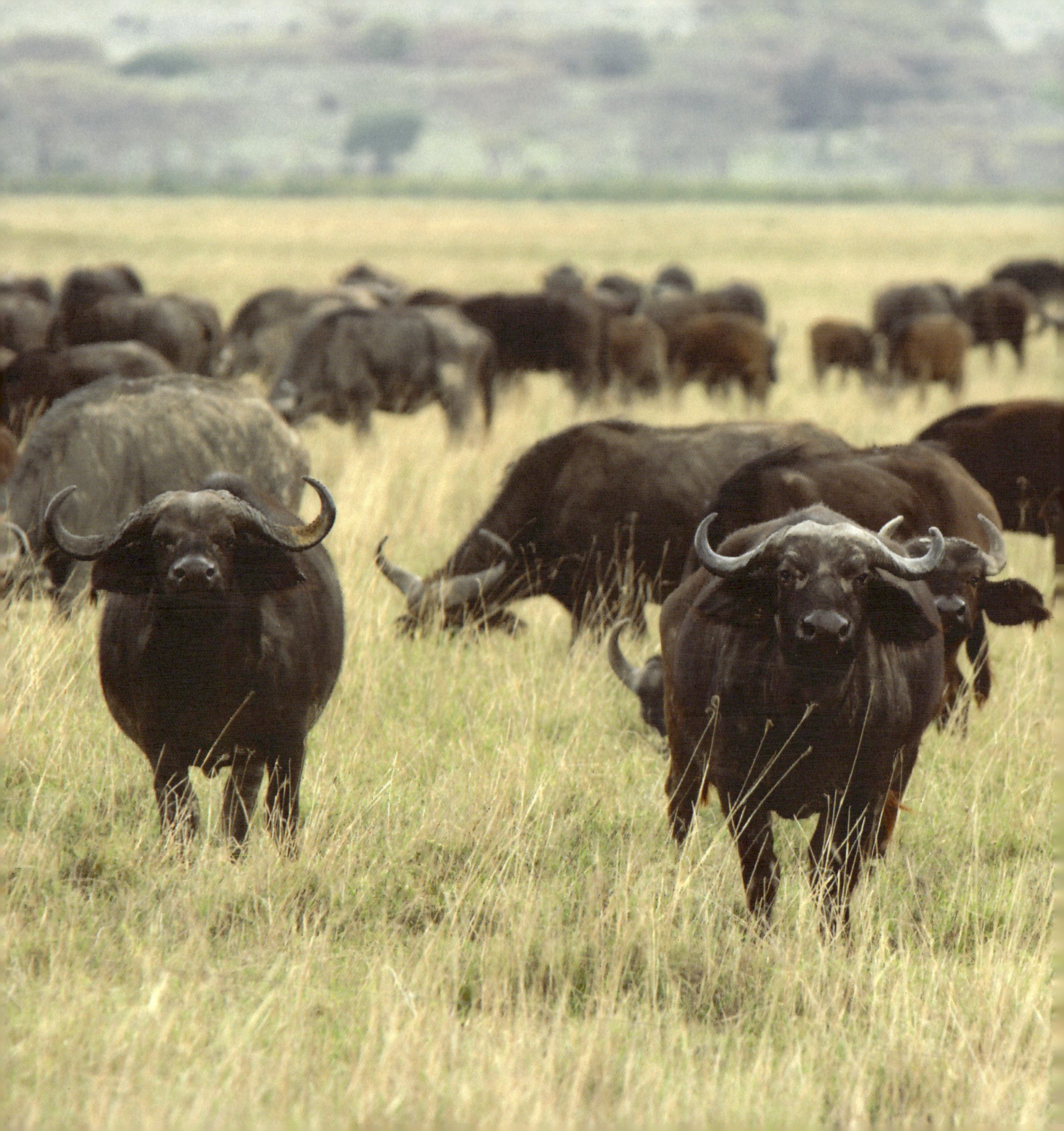

African buffalo has poor eyesight and sense of hearing, but their sense of smell is excellent.

Wildebeest is a mammal that belongs to the family of antelopes.

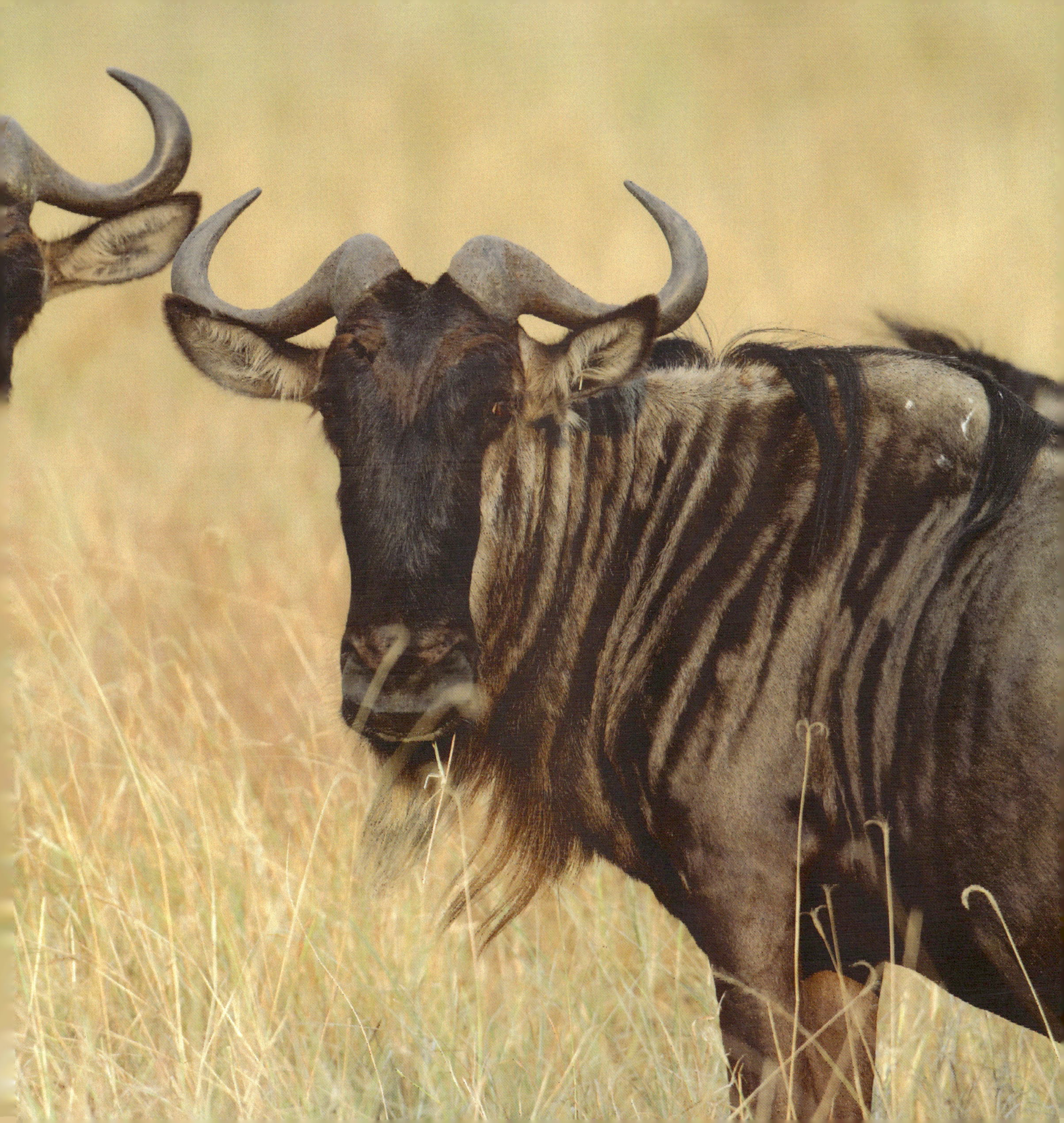

During migration, wildebeests travel between 500 and 1000 miles.

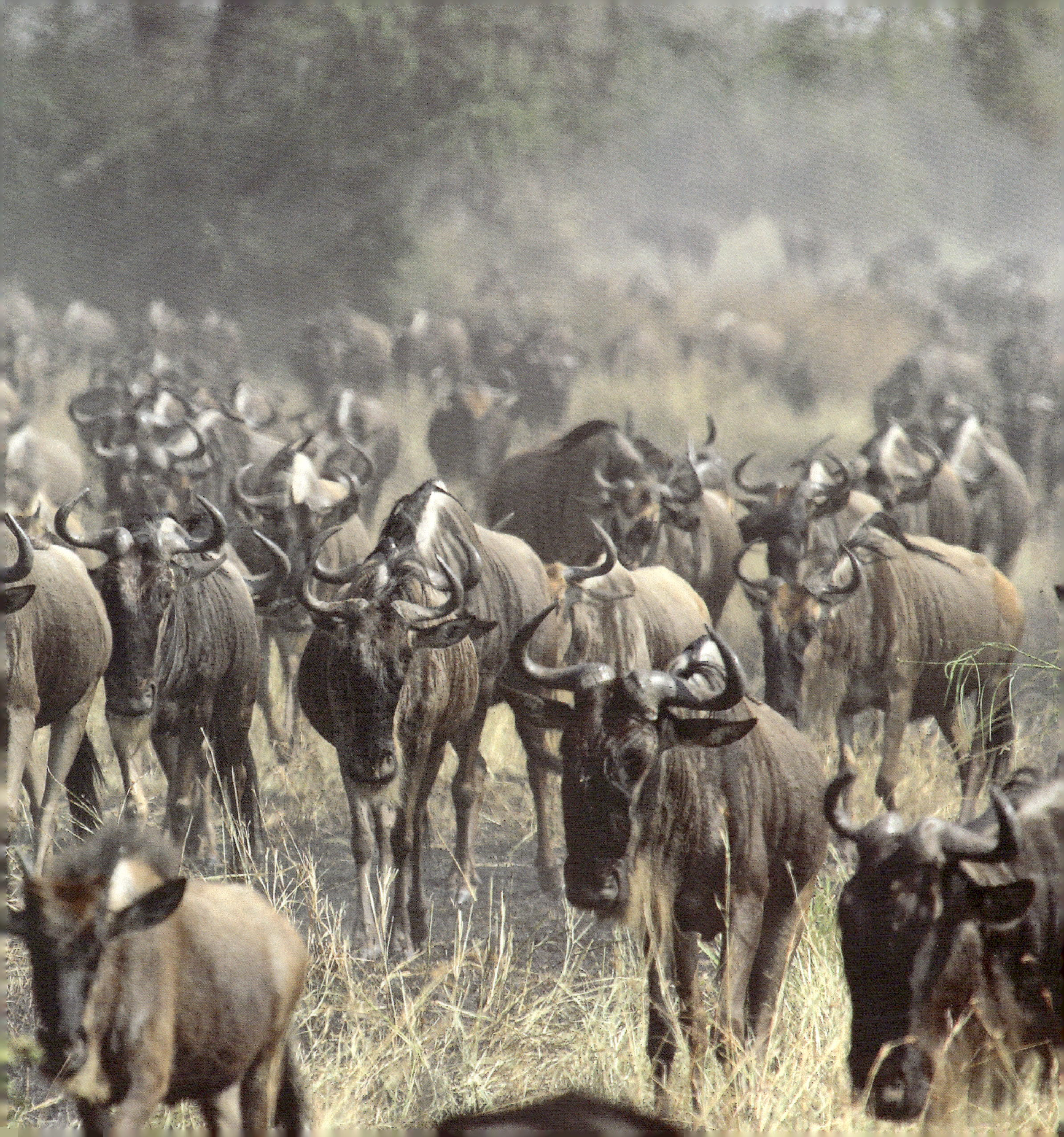

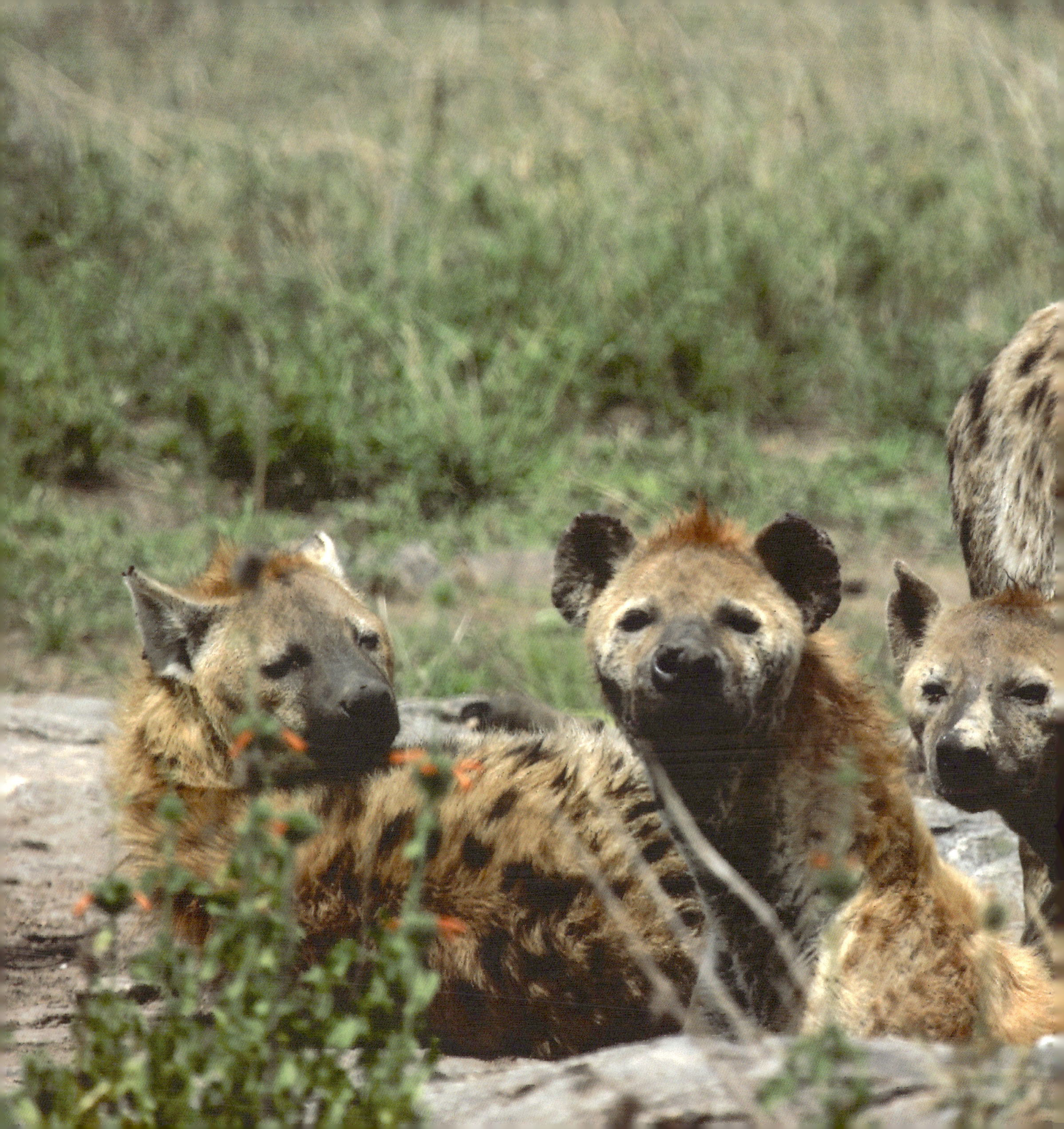

Hyenas are large, dog-like, carnivores. Hyenas use various sounds, postures and signals to communicate with each other.

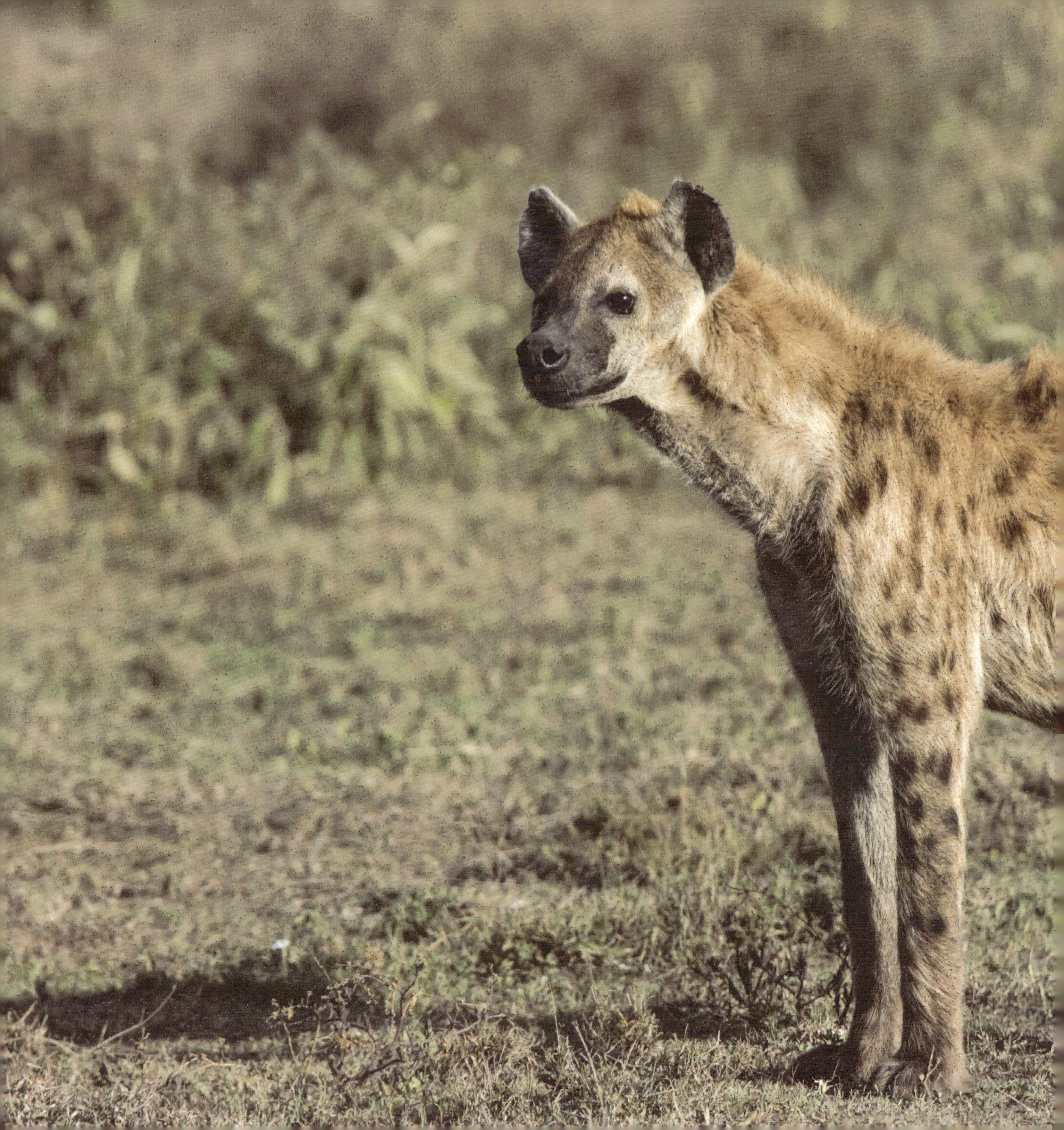

Hyenas live in territorial and large clans that can consist of up to 80 members.

Cheetah's are the fastest land mammal in the world. They can run at speeds up to 75 miles per hour.

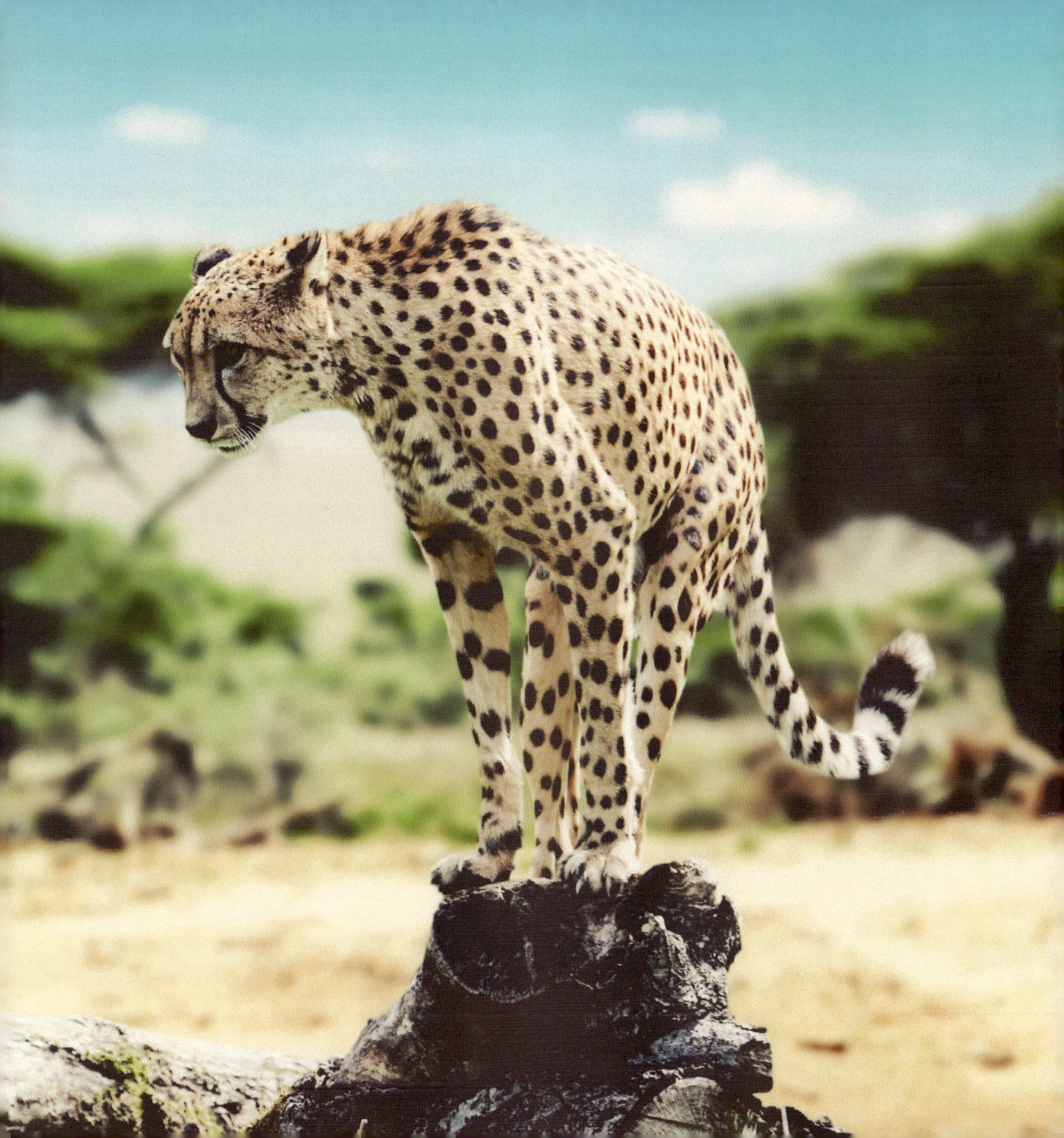

Cheetah's do not roar like lions or tigers. They let out a chirping noise when they feel threatened.

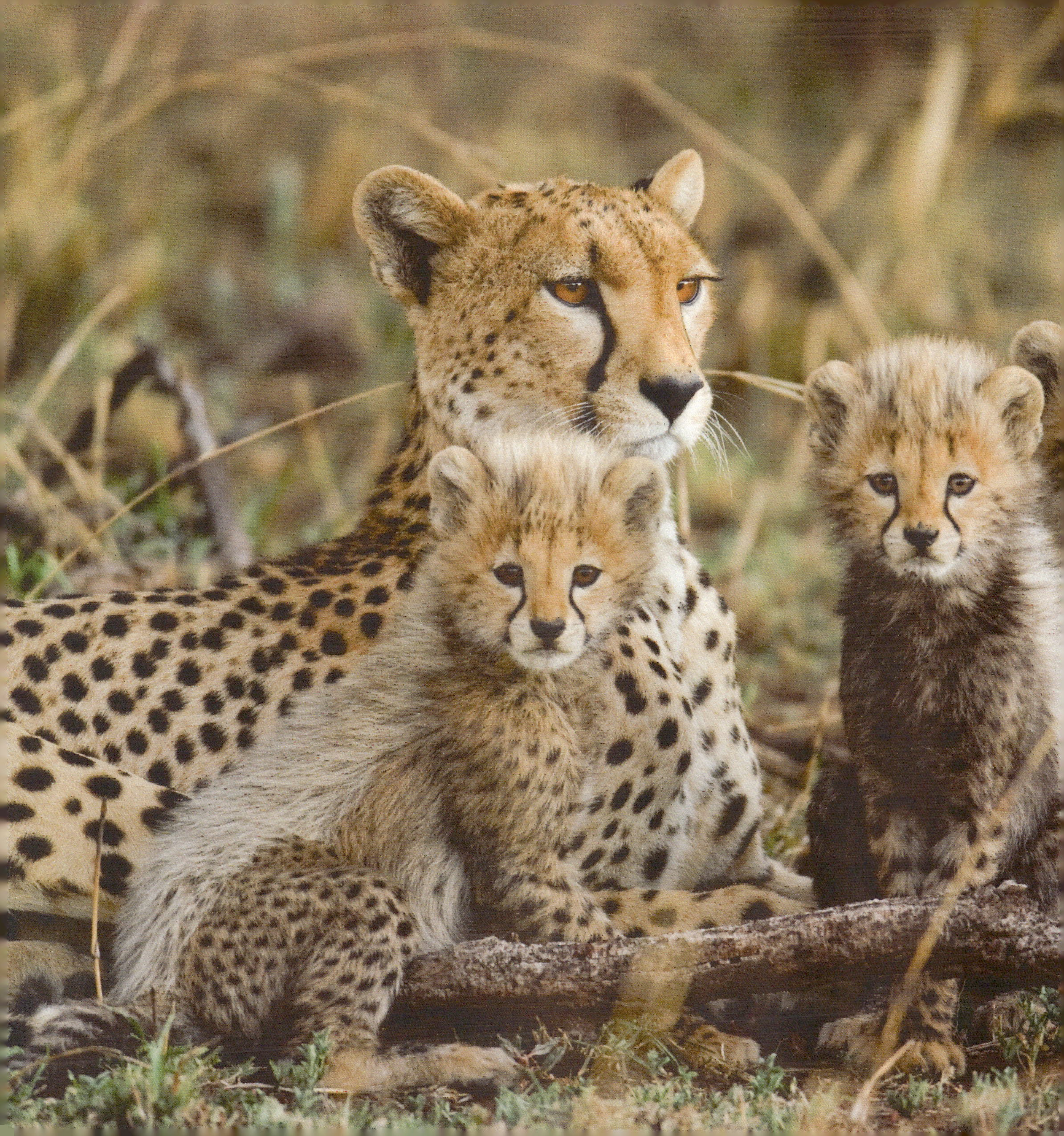

Printed in Great Britain
by Amazon